ANCHOR SYSTEM THINKING

The Art of Situational Analysis, Problem Solving, and Strategic Planning for Yourself, Your Organization, and Society

A. I. Shoukry

First published by A. I. Shoukry in 2018

Copyright © A. I. Shoukry, 2018

All rights reserved. No part of this publication may be reproduced, stored or transmitted in any form or by any means, electronic, mechanical, photocopying, recording, scanning, or otherwise without written permission from the publisher. It is illegal to copy this book, post it to a website, or distribute it by any other means without permission.

A. I. Shoukry asserts the moral right to be identified as the author of this work.

A. I. Shoukry has no responsibility for the persistence or accuracy of URLs for external or third-party Internet Websites referred to in this publication and does not guarantee that any content on such Websites is, or will remain, accurate or appropriate.

Designations used by companies to distinguish their products are often claimed as trademarks. All brand names and product names used in this book and on its cover are trade names, service marks, trademarks and registered trademarks of their respective owners. The publishers and the book are not associated with any product or vendor mentioned in this book. None of the companies referenced within the book have endorsed the book.

First edition

Paperback ISBN: 9781731199454

I dedicate this book to my wife, Heba

CONTENTS

Anchor System Thinking 1
Preface 6
Introduction 11
Chapter 1 14
Anchor System Thinking 14
    How the Mind Works 14
    How Anchors Work 17
    Anchor System Thinking Explained 21
Chapter 2 25
Anchor Points 25
    Time Anchor Points 26
    Location Anchor Points 34
    Social Network Anchor Points 39
    Entity Anchor Points 43
Chapter 3 49
Anchor Maps 49
    Personal Anchor Map 51
    Business Anchor Map 53
    Society Anchor Map 55
    Detailing 58
Chapter 4 61
Anchor System Thinking 61

for Problem Solving... 61
- Personal..62
- Organization 64
- Society 66

Chapter 5...68
Be the Anchor...68
Chapter 6 80
Proactive Anchoring..80
- Personal Proactive Anchor Map......................... 81
- Relationship Proactive Anchor Map....................84
- Organizational Proactive Anchor Map 86
- Society Proactive Anchor Map 89

Conclusion 91
Appendix..93
- Anchor System..93

Acknowledgements 96
About the Author..98
References 99

A. I. Shoukry

PREFACE

"I prepare all of my lectures and talks in my head while running," said my friend, as he watched me put headphones in my ears, prepare my music playlist, and get ready for our run together.

Yet, one Friday morning, I decided to give his words a chance while running alone on the beach. I felt free and was enjoying a moment of solitude. As I ran, a few questions surfaced in my mind: *"How free are we really are? And free from what? What things in life are we attached to?"*

Running alone, with no music playing, on an empty beach positively reflected a sense of freedom. If someone was standing on the hill and watching me run, it might remind them of a horse running along the shore, which is usually described as "free" and

ANCHOR SYSTEM THINKING

"wild." **But are we humans really free like those horses?**

Horses have foals, and mares worry about providing food and safety for them. But do mares and stallions worry about their foals when they are grown up? Do they think about their foals when they're running in the wild? Do they worry about their children while they're at work? Do they think about what their boss says as they're running? Do they have a boss or an assistant?

Of course, we can't know the answers to these questions. But we do know that us humans, even if we are running naked in the desert, will think about many things that keep us agitated or entertained. Even if you try to clear your mind of thoughts, you'll be thinking about the act of clearing your mind. It may only be a temporary clearance until you finish your run, a thought forgotten forever, or one soon replaced by other thoughts.

A. I. Shoukry

> *Our mind is a vacuum that always needs to be filled.*

While running one day, I noticed how I am anchored to many things surrounding me. Even when I try clear my head from work, I think about life, my family, or the books I want to write. I may enjoy the surrounding trees or clear desert—or watch out for reckless drivers while putting a thought to rest forever. Sometimes, I retrieve a memory of my home country Egypt and linger on it for a mile or two. Basically, there are three main processes going on in our brains: grasping new data, retrieving old data, and deleting old data.

Now, what makes certain data more important than other data? In other words, why is some data retained and other data deleted and forgotten? Why do I retrieve parts of these data sets while running? For example, thinking about my wife and kids, but not remembering other relationships with the same warmth and love? Why do I think about writing a book in the future? Why did I provide an example of a horse running and not a dog? And why do I think about Egypt, not New Zealand?

What I know is that I love my wife, my kids, my family, and my country. I am devoted to my work and my patients. I belong to my past, with all its memories, good and bad, and I aspire to be a better doctor and a writer. My thoughts and beliefs determine my decisions. But how can I describe all these relationships? Where do I stand in all of this? In an increasingly stressful world, the need to understand where we stand and where are we going is fundamental. Is there a map I can draw to know exactly where I stand? Is there a word, a formula, that can help me describe my position on the map and how to move forward?

Suddenly, as I was running and musing on this, the word "anchored" lit up in my head. Yes, I am anchored to all the things that surround me. My thoughts, memories, beloved ones, and work are all my anchors. I am the boat, life is the sea, and the

A. I. Shoukry

seabed is what the anchor holds on to. *Wow, I thought, this is a fresh idea I never thought I would have while running.* I guess my friend was right—run and listen to your mind.

INTRODUCTION

We are vessels in the sea of life. Whether we like it or not, whether we float or sink, stand still or move forward, we are always anchored or searching for an anchor. As I will explore in this book, we are either anchored to time, location, or social networks, and objective, subjective, or inter-subjective entities. (I will explain these complicated terms later.) Moreover, we may be anchored by multiple anchors at any one time.

The thing is, it's better to know your **anchor points** than to be anchored without knowing. If you think you are floating freely, think again. This is your introduction to **anchor system thinking**.

Anchor system thinking can be applied as a situational analysis tool on the personal, organizational, social, and even national levels. It is used to describe the current status and relationships in a meaningful and creative way.

Anchor system thinking can, therefore, be used to solve personal, business, political, societal, geopolitical, or relational dilemmas.

In this short book, you'll discover how to think using the anchor system concept, reach for an anchor, be an anchor for others, and use the anchor system for strategic planning for yourself, your organization, and society.

Chapter One: Introduces anchor system thinking and its application on the personal, organizational, and societal levels.
Chapter Two: Explains anchor points and how are we all anchored to location, time, social networks, and entities.
Chapter Three: Teaches you how to draw anchor maps for yourself, your organization, and society.

ANCHOR SYSTEM THINKING

Chapter Four: Shows you how to use the anchor system to solve personal, organizational, and social problems.

Chapter Five: Explains how to be an anchor point for others and how to excel at it.

Chapter Six: Shows you the proactive anchoring technique and how to use the anchor system for strategic future planning for yourself, your organization, and society.

A. I. Shoukry

CHAPTER 1

ANCHOR SYSTEM THINKING

HOW THE MIND WORKS

If our mind controls our body, then how does our mind work? The first part of this statement is the study of many scientific fields including neurology, physiology, anatomy, and others. The second part of the statement is the study of philosophers, behavioral scientists, and even theologians.

Our minds work more or less like a calculator. If you press 1 + 1, then press the equals key, what do you get? Obviously 2. If you press different numbers, you will get different answers, and equally, if you press a different function (such as the - sign), you will get another answer. This is just how our mind works: an input, a function, and an output. Now hold on to these three words: input – function – output.

Our senses are continuous sources of input, and everything we see, hear, smell, touch, and taste is an input. Such inputs are sometimes retained in our memory and other times discarded. Sometimes, an input is there, but we don't know that it exists.

The input is transformed through a complex neurophysiological process in our brains (function) to an (output). Functions can be DNA-prescribed, for example, when a baby's mouth touches its mother's nipples, the baby immediately suckles. They can also be learned and acquired functions, such as addition, multiplication, division, etc.

However, functions in humans, unlike calculators, may sometimes be altered, suppressed, or incited. Altered like they are in Alzheimer's disease, suppressed like when you suppress an instinct or will, and incited such as when you build on previously learned functions in your continuous process of education.

Then comes the output in the form of a memory, decision, behavior, or action.

The functions of our minds are affected by **anchors**. Anchors are external stimuli that result in imprints in our minds. They determine which data is retained in your memory, forgotten, or attempted to be forgotten.

This is why two people may walk together down the same street, but each remembers a different story of that street, although they have been exposed to the same experience.

In the next section, we'll explore how anchors work.

ANCHOR SYSTEM THINKING

HOW ANCHORS WORK

In 1904, the famous Russian Physiologist Ivan Pavlov won the Noble Prize in medicine. His work on classical conditioning is known as "the dog and bell experiment" and is very influential in the field of psychological therapy and the study of animal behavior. The basic idea was to associate an external stimulus with an internal response, in this case, a bell and the dog's salivation. At the start of the experiment, whenever food was introduced to the dog, a bell rang. Later, when the bell rang, the dog salivated—expecting food, but this time the food wasn't there.

Based on Pavlov's work, Neuro-linguistic Programming (NLP) coaches use this anchoring technique to associate external stimuli with an internal experience. Say you want to boost your confidence in public speaking. While preparing your lecture, you try to remember past experiences when you felt very confident about yourself.

Second, you associate a movement, such as clinching your fist or pressing your thumb and middle finger together. You then repeat these steps several times. As in the Pavlov experiment, when you need to remember the same confident feeling, later on, you repeat the movement you've trained yourself in. By relating the external stimuli to the internal feeling, you are anchoring yourself to a known point in the sea of the experience. While NLP has been heavily criticized as a theory due to a lack of scientific evidence, the word "anchoring" as they use it still describes a useful meaning—*reaching for something to anchor to.*

In psychology, there is the "anchoring bias" or "anchoring effect." This states that people tend to rely heavily on the first piece of information they find out about a subject. For example, when negotiating on a house, the first person who makes an offer (whether it's the buyer or the seller) will take the upper hand. The counteroffer is usually within close range of the first offer. Or if your parents didn't allow you to travel alone until you were 18 years old, when your kid asks you if they can go on a trip to Africa, you'll probably negotiate the same minimum age with them. You are anchored.

An anchor also means "principal support or something that serves to hold an object firmly."[1][2] Being anchored may mean "to be related to something," as in "this study is anchored to a theory" or "this man is anchored by his values."

A typical setting for the usage of anchors is with vessels in the sea. An anchor is a device generally made of metal with a chain that holds the vessel to the bottom of the sea and prevents it from drifting due to currents or the wind. An "anchor cable" or "anchor chain" is used in larger vessels and is made of chain, cable, rope, or a combination of these. A vessel usually carries one or more anchors on board, which are considered temporary, while permanent anchors are left in a fixed location and are used for mooring.[3]

Anchors achieve holding power either by "hooking" into the seabed, holding a boat in place by sheer mass, or a combination of the two. Permanent moorings use large masses such as a block of concrete resting on the seabed. Anchors on large ships derive a significant portion of their holding power from their mass in addition to hooking to the seabed. Smaller vessels have metal "flukes," which hook onto rocks on the bottom or bury themselves in the soft seabed.[3]

ANCHOR SYSTEM THINKING EXPLAINED

When the word "anchored" popped into my mind while running, it was based on the typical usage of anchors by sea vessels. A vessel is held by a chain, which ends with an anchor secured to the seabed. I wondered, can this concept and terminology be used to describe how we as humans are anchored to life? And can it be further extended to cover larger-scale entities? The answer is yes.

This is what I call **anchor system thinking**. It is a holistic approach of thinking based on the sea anchors metaphor. Our anchors are mind imprints that are innate or result from external stimuli, and these anchors are points that hold us in a drifting sea (life). When we receive input from our senses, retrieve a thought from our memory, or try to delete a memory, anchor points affect how the data is processed (function) in our mind. This results in a change in our behaviors, actions, and decision-making.

By identifying our anchor points, we can understand where we stand (anchor maps) and can proactively plan where we are going. Just like we hold to anchor points, we are sometime anchor points for others.

Anchor system thinking is not only used on the personal level, but it can also be applied on organizational and societal level for situational analysis, problem-solving, and strategic planning.

The anchor system is formed of five components, which will be explained in the following section:

- **The vessel**
- **The medium**
- **The anchor**
- **The bed**
- **The chain**

ANCHOR SYSTEM THINKING

The vessel is tied with a chain and anchor to the bed of the medium. This is the five-component form of the anchor system. Let's see how this system can be applied to the personal, organizational, and societal levels.

The vessel

The vessel can be a person such as yourself, a friend, your husband/wife, etc. Or it can be an organization such as a company, charity organization, political party, etc. It can also be a nation, country, culture, or religion, etc.

The medium

When using the anchor system on a personal level, the medium is life. If the anchor system is applied to solve a marriage problem between a husband and wife or a couple planning a long-term relationship, the medium may be a marriage. When planning a campaign for political parties, the medium would be the political sphere. The medium for business plans for start-up companies or long-term planning for established ones is the market. To solve diplomatic relations between two or more countries, the medium would be geopolitics. To understand the rise, spread, or fall of religion in certain countries, the medium would be society.

The anchor point

The vessel is connected by a chain that ends with an anchor that hooks to the bed of the medium. The anchor (A), bed (B), and chain (C) are the ABC, and together they form the **anchor point**. We'll explore anchor points in the following chapter.

CHAPTER 2
ANCHOR POINTS

As you saw in the previous chapter, the anchor (A), bed (B), and the chain (C) are the ABC that together form the anchor point. Everyone and almost everything in this world are anchored. Every vessel in every medium is anchored to one or more anchor points. There are four types of anchor points:

- Time anchor points
- Location anchor points
- Social network anchor points
- Entity anchor points

A. I. Shoukry

TIME ANCHOR POINTS

On the personal level, you can be anchored to the past, present, future, or a combination of these **times**. For example, next February, I am planning to participate in the first marathon to be held at the Pyramids of Giza, so I got a pen and paper, calculated the number of weeks left to race day, and planned my training program. I anchored myself to the future.

Getting anchored to the future means you are optimistic, moving forward, and a visionary. But getting anchored to the future without acting in the present just makes you a daydreamer. I need to train and put some miles in the bucket, so I'm ready on race day.

When I wake up on work days, I check my calendar to see what type of surgery I have that day, and I am anchored to the future. However, while performing the surgery, I am anchored to the

ANCHOR SYSTEM THINKING

present. I cannot be somewhere else, or I will lose concentration and might put my patient and myself in trouble.

Some people say, "I live in the present. I don't look back or forward. I just enjoy the moment." That's good but living in the present means you're only anchored in the present. Knowing this is important—to understand the meaning of your life, what direction you're going in, and how you spend your time. However, being anchored to the present does not mean you shouldn't or couldn't be anchored to the past and future too.

> *We are all anchored to all-time tenses.*

Other people are anchored only to the past, which is not sustainable. They keep saying "when we were young" and "back in the good old days," as if the present or future is always evil. However, you cannot only be anchored to the past. I enjoy reading the history of Egypt as I like to enforce my anchors to the past, but I am anchored to the future and present too. While discussing the anchor system with one of my friends, he described the good old days when he used to visit his beloved late grandmother. He kept telling me detailed stories about her delicious cookies, bedtime stories, and strong character. I told him promptly, "You are anchored to your past, my friend."

Similarly, imagine a single mom whose husband cheated on her, who cannot start a fresh, new relationship with other men. She is anchored to the past in a way that she is unable to move forward. By contrast, imagine a woman who is the CEO of an explosively-growing IT start-up. She remembers the early eighties when she used to sit beside her father as he programmed for hours on the DOS IBM machines. He taught her basic concepts of coding using punch cards. She remembers precisely how she always dreamed of working on the actual keyboard, and her best day ever—when she did her first "Hello World" program. She is anchored to the past in a way that lit a fire to her future.

ANCHOR SYSTEM THINKING

You can be anchored to all-time tenses. The truth is that we are all anchored to all-time tenses whether we like or not—it's just that many of us do not recognize it.

An organization, such as clubs, charities, or business companies, may also have time anchor points. For example, the famous Swedish car company Volvo has a mission for the future of the company:[4]

> *To be the most desired and successful transport solution provider in the world.*

In fact, every mission for any company or organization is an anchor to the future. But anchors can cover a wider scope. For example, Volvo has announced that 50% of their car sales will be fully electric by 2025, a precise goal that anchors them to the future. But the same company may be anchored to its past by its brand heritage, just as Volvo cars are known for their safety and reliability.

> *Every mission for any company is a time anchor point to the future.*

Time anchor points can be used to describe a nation's history and its civilizations. Although the Pyramids of Giza are the most famous pyramids in the world, they aren't the first built in Egypt, nor are they the only pyramids. The Pyramid of Djoser was built around 50 years before them, and Egypt has more than 100 pyramids. Egypt is anchored to its past by the Pharaonic civilization. At the entrance of the Grand Egyptian Museum, the mega-construction project stands the statue of Ramses II. However, Egypt is also anchored to its Roman, Ptolemaic, Coptic, Arab, and Muslim heritage. Each one of these forms a separate time

ANCHOR SYSTEM THINKING

anchor point that can be illustrated, described, and discussed.

You can't open a history book on the United States of America without discussing the immigration of the first settlers. Other than the small percentage of Native Americans, all Americans can trace their ancestry to other nations around the world. This is an important time anchor point for the United States.

By the end of January each year, the World Economic Forum held in Davos is a present time anchor point for Davos town. All current events are present time anchor points. In fact, the news is a collection of bullets of time anchor points for each city, country, and people.

> *Current events are simply present time anchor points.*

Future time anchor points for countries are cornerstones for planning and moving forward.

Ministries, political parties, and parliamentarians discuss missions, visions, plans, and programs, and these are all future anchor points for countries and cities.

On the societal and cultural levels, time anchor points are crucial in understanding the origins of religions, cults, and movements. The history, the early days, the rise, and the fall. All religions are anchored to multiple time anchor points. The crucifixion of Jesus for Christians and the migration (Hijrah) of Muhammed from Mecca to Medina are foundational past time anchor points for Christianity and Islam respectively. Pope visits, Ramadan, and Haj, are all present time anchor points. The Oscars, Grammy Awards, and Cannes Festival are all famous present time anchor points for the entertainment industry.

ANCHOR SYSTEM THINKING

There is no better way to explain social and cultural future time anchor points than UNESCO's mission: "to contribute to the building of peace, the eradication of poverty, sustainable development, and intercultural dialogue through education, the sciences, culture, communication, and information."

A. I. Shoukry

LOCATION ANCHOR POINTS

The second type of anchor point is **location**. Every one of us has a place of birth. We grow up, and while some of us spend the rest of our lives in the same neighborhood, others leave their community or country for good. Whether we like it or not, we can't escape the location anchor. If we love our country, city, or neighborhood, then we anchor ourselves with a chain of love. However, if we despise the place, then it's a chain of hate.

Under various circumstances, people are forced to leave their country of birth and become citizens of a new country. Some people easily integrate with their new communities, while others fail to find a new anchor and instead keep their old chain and anchor tied to their home country.

Getting anchored to places is not just limited to countries, cities, and neighborhoods either. People get anchored to their workplaces, schools, colleges, or homes.

Sanjay and Manoj are two friends who went to the same high school in India, were raised in the same neighborhood, and always dreamed of working as doctors in the USA. Both finished medical school and got very high scores in the USMLE exams, then applied for residency in the States. After completing their residency, they finally achieved their childhood dream. However, while Manoj was happily married and living the American dream to the full extent, Sanjay never stopped thinking about going back to India. It's not that Manoj didn't love his home country, but Manoj managed to set a new location anchor point in the USA as well as his old location anchor point of India. Sanjay, however, failed to form a new location anchor point in the USA, and his old location anchor point of India kept pulling him back. Sanjay finally made the right decision for him—he went back to work as a consultant in his hometown hospital.

A. I. Shoukry

On the organizational level, a company is anchored by the location of its headquarters and its branches. When an operation manager needs to decide where to distribute a product, he is anchored by the storage and distribution centers.

When Amazon announced that it would build a second headquarters, a project named HQ2, 238 locations from Canada to Mexico submitted bids.[5] The project is expected to provide employment for 50,000 people. As such, Amazon's criteria for the location included a business-friendly environment and a metropolitan area with more than a million residents. Amazon precisely understands the concept of anchors. The choice of location anchor point is a fundamental decision that many future decisions will depend, for Amazon and anyone. It will anchor the company for years to come and may decide its future.

ANCHOR SYSTEM THINKING

A country is anchored by its location; it cannot change its location. It usually has fixed borders with other countries. However, with the opening of the Suez Canal, constructed between 1859 and 1869, the international water navigation changed forever, and with it, its geopolitical location.[6] The artificial waterway built in Egypt offered a shorter journey for vessels between the North Atlantic and northern Indian Oceans via the Mediterranean Sea and the Red Sea. By avoiding the South Atlantic and southern Indian Oceans, it reduced the journey of commercial ship carriers by approximately 7,000 kilometers. The Suez Canal opened a flood of foreign currency income to Egypt. The creative thinking of changing the geopolitical location anchor point of Egypt by constructing the Suez Canal had an economic and political impact on the Middle East and world history that lasted forever. It gave Egypt an economic and political advantage by controlling one of the most strategic navigation canals in the world.

Societies and cultures are also anchored by location anchor points. New York City is a mecca for people who love theater, and Las Vegas is a mecca for gamblers. The Vatican City is a location anchor point for the Catholic Church and for more than 1.3 billion people around the world. Muslims are anchored by Mecca, and every year at the same time on their Hijri Calendar, they visit Mecca to perform Haj, Islam's fifth pillar. They don't go to Cairo or Dubai—they go to Mecca. When Muslims pray, they pray towards Mecca. It's their location anchor point.

ANCHOR SYSTEM THINKING

SOCIAL NETWORK ANCHOR POINTS

Our parents, biological or non-biological, are our first **social network anchor points** in life. When we grow up, we get more anchors: siblings, friends, teachers, or even neighbors. Then comes our extended family, work, church, or mosque groups.

Popular social network anchor points are our children, love partners, and parents. Loving partners may be our solid anchors in tough times, or sadly sometimes the anchors we don't find when shit happens. For many people, children are their sole anchors to stay alive for.

Social network anchor points, as the name implies, usually come in numbers, and involve a group of people and networks. Your parents, love partner, children, and siblings can each be a social network anchor point. And together, they might form a family social network anchor point. Your friend Mike may be one social network anchor point, but all your friends together including Mike may form a friend's social network anchor point. It's you who decides who is included in your social network anchor points and who is not.

On the organizational level, a business company is anchored by its human resources, while all the management, staff, and personnel are social network anchor points. They might also be called internal social network anchor points. Suppliers, distributors, and outsourced manufacturers are all social network anchor points. They might also be called external social network anchor points. And because the social network anchor points are used to describe relations with the "other," competitors and sister companies are also social network anchor points. For example, *The New York Times* is anchored by its journalists and writers internally, and by freelancers and distributors externally, and

ANCHOR SYSTEM THINKING

by other newspapers such as the *Washington Post* and the *Financial Times*.

Social network anchor points for a country are its diplomatic relations with other countries, where each country forms a specific social network anchor point. For the Ministry of Investment in any country, foreign and local investors form its social network anchor points. For a Tourism Ministry, travel agencies, tourism companies, and delegations are social network anchor points. A country is also anchored to the people living inside it, and by the expatriates living abroad. They all form different social network anchor points.

At the social and cultural levels, social network anchor points are usually the founding fathers of ideas, opinion leaders, principal spokesmen, etc. In any religion, social network anchor points are prophets and messengers. In capitalism, the social network anchor point is Adam Smith, and in Marxism, it is Karl Marx.

ANCHOR SYSTEM THINKING

ENTITY ANCHOR POINTS

So far, we have identified different anchor points for time, location, and social networks. But what about the other things we see, hear, feel, or smell? What about the other things we think about, like love? What about God, Google, and the nation? These are **entity anchor points**.

These anchor points are classified into objective, subjective, and inter-subjective entity anchor points. **Objective entities** are objects outside the subject (ourselves), such as trees, cars, and houses. They all exist in real life—we can see, hear, and smell them. **Subjective** entities exist in your own subjective conscious. They are individualistic experiences that can be described to others but not witnessed by others, like your child's imaginary friend. Then there are **inter-subjective entities**, which exist in the collective consciousness of the group and are shared by subjects, such as laws, corporations, nations, and God.

A. I. Shoukry

In my book *In or Out: A Practical Guide to Decision Making*, I explore the factors that influence our decision-making process and categorize them into **internal factors** and **external factors**.

Internal factors include both hard and soft factors. Hard internal factors play a part in the decisions we make and cannot be easily altered, such as how our brain is built to work—our cognitive systems. This includes our morals, core values, and principles.

Soft internal factors are easier to change and are often influenced by our circumstances at the time of making the decision. These include intentions, interests, harms, objectives, and goals.

Both hard and soft internal factors that influence your decision-making process are subjective entities. On the personal level, my subjective entity anchor points would be my values, principles, and goals.

On the organizational level, Mark Zuckerberg outlines five core values for Facebook: focus on impact, move fast, be bold, be open, and build social value. These five values and the company's internal code of ethics are some of Facebook's subjective entity anchor points.

On the societal level, it's tricky to distinguish subjective from inter-subjective entities. If the entity is shared only by a particular society and not by other societies, such as the USA but no other countries, then the society is considered a whole and treated as one subject, as an individual. As such, a society's customs and traditions would be subjective entities, and society is anchored by subjective entity anchor points.

Objective entities are much easier to explain. They can be seen, heard, touched, or smelled by anyone. A car is a car, a tree is a tree, and whether you are young or old, Chinese or American, we all agree on what a car and a tree are. Objective entities can usually be scientifically proven to exist.

In my book *In or Out*, I also explore external factors that influence the decision-making process. External factors are those outside our body that determine or constrain decision-making, and they include resources.

ANCHOR SYSTEM THINKING

On the personal level, resources such as my car and my apartment are objective entities. On the organizational level, company resources such as products and machines are objective entities. On the societal and cultural level, mosques, church, musical instruments, pens, and paper are objective entities. All can form objective entity anchor points.

Inter-subjective entities are shared by subjects' collective consciousness. They might not exist in real life or be scientifically proven, and their value is based on the wide, collective acceptance of the story of their existence. The more people trust and believe in their value, importance, and existence, then the higher their acceptance and value. Money is a famous example. On the personal, organizational, and societal value, money is an inter-subjective entity anchor point.

For a business company, country and market laws are inter-subjective entity anchor points. For countries, international laws and international organizations, such the United Nations and NATO, are inter-subjective entity anchor points.

On the society level, the Quran and the Bible are inter-subjective entity anchor points for Islam and Christianity respectively. Yoga is an inter-subjective entity anchor point for India and Hinduism.

To wrap up this chapter, we have a vessel, a medium, and anchor points. Anchor points can either be time, location, social networks, or entities. In the following chapter, you'll see how to use the anchor system as a situational analysis tool to understand your **current anchor status**.

CHAPTER 3
ANCHOR MAPS

A boat that embarks from point A to point B leaves the mooring to anchor with another. If we can draw a map of how this vessel moves in the sea, where it embarks, when it moves, and why it moves from point to point, we could gather much information, and from this, we could improve the boat's pathways, timings, and movement.

An **anchor map** would do this job perfectly. An anchor map is a drawing that illustrates a vessel in a medium and all of its anchor points. A finished anchor map will show the current anchor status of the vessel.

Before drawing an anchor map, here is a suggested map key. A blank piece of paper is the medium.

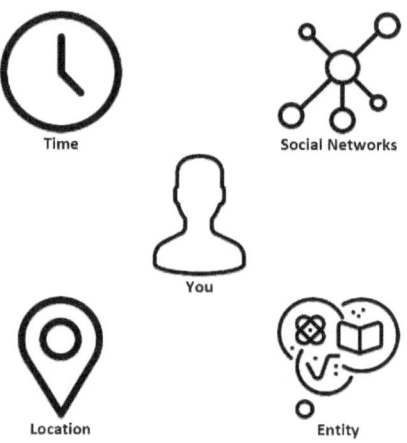

Next, you'll see three anchor map examples. The first is a personal anchor map, and it shows you how anchor maps can be used to draw the current anchor status for an individual. The second anchor map is for a business company, and you'll see how anchor maps can be used to illustrate the current anchor status for an organization. The third example will show you how to use anchor maps to understand society.

ANCHOR SYSTEM THINKING

PERSONAL ANCHOR MAP

On the personal level, we are vessels in the sea of life, and we are always anchored to at least one—if not many anchor points. It's better to know your anchor points than to be anchored without knowing.

Grab a pen and paper and start drawing:
Draw yourself at the center of the paper.
Define your:
- *Time anchor points*: Which anchor points hold you to your past, present, and future?
- *Location anchor points*: Your home, city, workplace, or country?
- *Social network anchor points*: Partner, children, friends, or any social network you belong to, such as your club, fraternity, church, religious group, etc.
- *Entity anchor points*: What are your objective, subjective, and inter-subjective anchor points?

Improve: Now decide which anchor points are adding value to your life and need to be fostered, and which anchor points are holding you back and need to be deleted or fixed.

Below is an example of a personal anchor map.

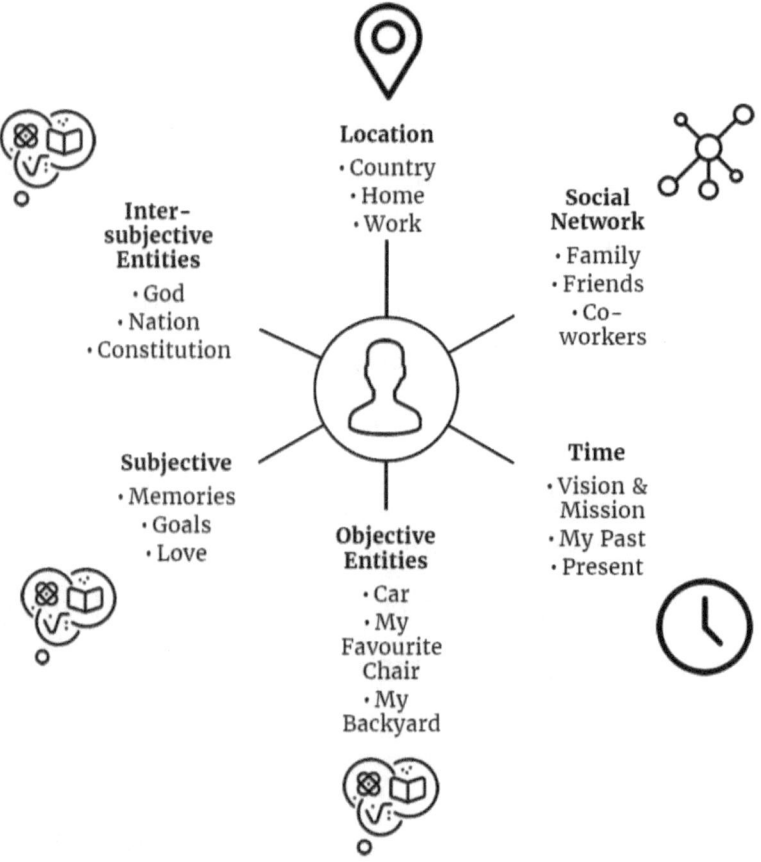

ANCHOR SYSTEM THINKING

BUSINESS ANCHOR MAP

On the organizational level, anchor maps can be used to illustrate the current anchor status of any organization, such as a business company. If you have a start-up, you can draw the map yourself, or if you are the CEO of a large company, you may ask your strategic development department to draw up an anchor map for your company.

Now imagine you own a small online business, and you'd like to have an anchor map for it. Grab a pen and paper and start drawing:

Draw your .COM business at the center of the paper.

Define your:

- *Time anchor points*: Which anchor points hold you to your past, present, and future? What are your company goals? What is your company mission and vision?

- *Location anchor points*: Your physical business address, hosting DNS and server, mirror servers, etc.
- *Social network anchor points*: Your sister websites, competitors, suppliers, distributors, content editors and creators, designers, programmers, etc.
- *Entity anchor points*: What are your objective, subjective, and inter-subjective anchor points? What are your resources, code of conduct, terms and policies, etc.?

Improve: When you have your current anchor status, decide which anchor points are adding value to your start-up and need to be fostered, and which anchor points are holding you back and need to be deleted or fixed.

ANCHOR SYSTEM THINKING

SOCIETY ANCHOR MAP

Anchor maps are used to understand the relationships of a vessel or multiple vessels in any medium. Anchor maps may be used to study societies, people, clubs, religions, or cults. Next, you'll find an example that studies the rise of a cult in a society, which is useful for marketers and sociologists alike.

Have the icon for CultX at the center of the paper.
Define the anchor points:
- *Time anchor points*: Which anchor points hold CultX to the past, present, and future? What is its history? What are its goals, mission, and vision?
- *Location anchor points*: The cult's city of origin, active cities, methods of spread, etc.
- *Social network anchor points*: Cult key figures, influencers, spokesmen, opinion leaders, opposition figures, cult prophets, etc.

- *Entity anchor points*: What are your objective, subjective, and inter-subjective anchor points? What are their resources, code of conduct, values, and principles, etc.?

Understand: When you've illustrated the current anchor status, understand which anchor points are adding value to the cult and which aren't. For example, the game *Fortnite* became a cult among teenagers very quickly. Epic Games, the parent company, can identify which celebrities, influencers, and famous online gamers play the game. These are social network anchor points. By visiting each anchor point, strategist, marketers, and leaders can decide what the strengths and weaknesses of the cult are.

Act: This is when the marketing department steps in. In the *Fortnite* case, they can now enforce a social network anchor point for the game and start marketing and advertising.

ANCHOR SYSTEM THINKING

Decide: Depending on your position, this is where decide which anchor points need to be fostered and which anchor points need to be deleted or fixed.

DETAILING

Optionally, you can make your anchor maps more advanced by adding more detail to them. Start thinking deeper about each anchor point. For each point, you need to identify (if possible):

The anchor: Permanent or temporary?
- A permanent anchor is one that cannot be removed.
- A temporary anchor is one that can be removed.

The bed: Fixed or dynamic?
- A fixed bed is a passive bed. Only you can anchor and remove the anchor—it's a one-way relationship.
- A dynamic bed is an active bed. If you can remove your anchor from your side whenever you decide to, so can the bed. If you are anchored to your wife or work, both can cut the anchor chain loose if they decide to do so, by leaving or

firing you respectively. Remember, anchors in personal relationships are usually dynamic.

The chain: Material and length?
- *Chain material*:
    - Made of steel, is hard, and cannot easily be removed or cut.
    - Made of rope, is soft, and can easily be removed or cut.
- *Chain length*:
    - Fixed length—does not allow maneuvering.
    - Flexible length—does allow maneuvering.

So, we've looked at how to use anchor system thinking as a situational analysis tool and use anchor maps to illustrate the current anchor status on the personal, organizational, and entity levels.

A. I. Shoukry

In the following chapter, you'll learn how to use the anchor system for problem-solving.

CHAPTER 4

ANCHOR SYSTEM THINKING FOR PROBLEM SOLVING

Anchor system thinking can be used to solve personal, organizational, and social problems. As such, the concept can be applied to solve business, political, geopolitical, or any relational dilemmas.

In any relationship, you need to identify the participating vessels and their common medium, then start by defining the anchor points.

PERSONAL

On a personal level, the anchor system can be used to solve relationship problems between a husband and wife, two friends, children and parents, siblings, etc.

Let's explore a problem between a father and a daughter:

Vessel
- Father
- Daughter

Medium
- The family

The anchor map
- Identify the anchor points that hold both vessels (father and daughter) together.
- Identify which anchor points they share, and that can be enforced.

ANCHOR SYSTEM THINKING

- Identify which anchor points that hold each of them separately can be shared, weakened, or deleted to have a better relationship.
- Identify what anchor points they can newly create to hold them together in a stronger way.
- Explore all time, location, social network, and entity anchor points.
- Draw a new anchor map that they share.

ORGANIZATION

The anchor system can also be used to solve problems on the organizational level. For example, when two companies are negotiating a merger.

Vessels
- Company A
- Company B

Medium
- The market

The anchor Map
- Identify the anchor points that hold both vessels (companies) together.
- Identify which anchor points they can work on to complete the merger.
- Identify which anchor points hold each vessel separately and can be shared after the merger to save resources, decrease costs, maximize profits, etc. Or which anchor points can be deleted or weakened to complete the merger,

ANCHOR SYSTEM THINKING

for example, some workers may need to be laid off.
- Identify which anchor points they can newly create after merging, such as a new headquarters.
- Explore all time, location, social network, and entity anchor points.
- Draw a new anchor map for the merger.

SOCIETY

Likewise, it can solve problems on a societal level. Say the Foreign Affairs Ministry is working on resolving relations between their country and its neighbor country.

Vessel
- Country A (our country)
- Country B (foreign country)

Medium
- International relations

The anchor map
- Identify the anchor points that country A and country B share.
- Identify which anchor points country A can work on to enforce relations.
- Identify which anchor points the countries don't share.
- Identify which anchor points country A should start to work on to share with country B.

- Identify which anchor points country A should cut or weaken to improve relations with country B.
- Identify which anchor points country B has that country A can work on to improve their relationship.
- Identify what new anchor points country A can create to share with country B.
- Explore all time, location, social network, and entity anchor points.
- Draw a new anchor map for the two countries.

In the following chapter, you'll learn how to be an anchor point and excel at it.

A. I. Shoukry

CHAPTER 5

BE THE ANCHOR

As people may be anchor points for us, we too can be anchor points for others. Sometimes we are anchors, and we know it, and other times we do not. Sometimes, we think we are an anchor, but we're not.

When your boss puts a hand on your shoulder and tells you, "I will depend on you during this mission," they are basically anchoring you. You are a social network anchor point for your company on this mission. If your spouse, children, or parents draw their anchor map, you expect to be one of their social network anchor points. If you're running for office or serving in any public duty, you are an anchor point for the people. A soldier in the field is an anchor for his nation, and an Imam in the mosque or a father in a church are both anchors for their religions. You are an anchor for your dog, and they are an anchor for you.

If you are an anchor point and you know it, you should genuinely act as an anchor point. An anchor that anyone can really depend on when it's needed.

Conversely, problems may occur when you don't know that you are an anchor for others, or when you think you are an anchor but you're not. However, if you act genuinely as an anchor point, you will never be in this situation—or if you're already in this situation, you'll know how to become a better anchor.

So, let's look at how you can excel as an anchor point for others.

Awareness

Know your anchor points.

First things first, you should be aware of your surrounding social networks, what role you play in each of them, and what your duties and responsibilities are. Who expects you to act as an anchor point for them, and to whom you expect to be anchor point?

Grab a pen and paper and start drawing.
Start from the inside, with the closest circuit, and move outward.
Usually, your family is the first circle. What roles do you play there? Husband/wife, child, sibling, father/mother?
The second circle could be your work, neighborhood, or charity organization.
Then there are the broader circles, such as your nation, religion, or political beliefs.

For each circle, know who is expecting you to be an anchor point and which anchor points you want to stand for.

ANCHOR SYSTEM THINKING

Communicate

> *Communicate, as people need affirmation and reassurance.*

After you have listed the anchor points you want to stand for, don't leave any space for failed expectations or miscommunication. Go and communicate with the people in the circles where you should and expect to be an anchor point.

Some anchor points are taken for granted, for example, you are expected, due to responsibility and love, to be an anchor point for your spouse and kids without needing to clarify this. However, it is of utmost importance that you emphasize that you're there and available to them. Sometimes, people need affirmation of this. They need reassurance, to feel safe, and to make sure that when they fall, you will be there for them. So, go and tell your family and friends that you are there for them and are always going to be there with help, care, and love.

At your work, communicate with your boss and co-workers and assure them that you are here for the company and are willing to give it your all when they need you.

Available

> *To be an anchor, you must be there.*

It's not enough to just talk though. Talk without the walk ends in disaster. If you have decided to be an anchor point, you should be responsible—as failure to show up when you're needed and expected is far worse than turning up when you're not expected.

To be an anchor, you must be available for other vessels that depend on you. Be reachable, and answer when they call for help. If you're going to be unreachable or can't be available, for example, you are traveling abroad or entering an area with weak cellular signal, tell others how to reach you and what to do if they can't contact you. Prepare a backup in case of emergency, maybe a second person on call, who will fill in for you if someone can't reach you.

Informed

> *Be proactive and always make the first move.*

Always stay informed and up to date with the latest news for all the people and entities who are anchored to you.

Don't wait for them to contact you—be proactive and always make the first move. I like to stick to the habit of eating lunch with my wife and kids whenever possible, so we can all exchange the latest updates on our lives.

At work, it's important to have social gatherings every so often to get closer to your co-workers and avoid miscommunication, as this can happen due to infrequent or lack of casual communication.

Listen

Good listening is better than great talking.

Being an anchor point doesn't mean you are a mentor or playing a teacher role. You could be an anchor point for your parents, friends, or co-workers. Sometimes, just being in the right place to care and listen is the expected role of an anchor point. Listening to others is crucial when being a social network anchor point.

Knowledgeable

Update yourself on the latest knowledge in the field of your anchor point.

You are an anchor point for others because they need the knowledge you have or the value you provide. Either they can't get it elsewhere, or you offer it at a lower cost, with better value, or in a unique way. Sometimes, you are their only source of information. So always update yourself on the latest knowledge in the field of your anchor point, be it parenting, work, the history of your country, etc.

Being knowledgeable doesn't need to be serious, for example, I like to keep up to date on the latest TV shows, movies, and music that my kids like, so I can always communicate with them and speak the same language.

Resourceful

> *Actively engage with the problem.*

People reach out for their anchors to help them in difficult situations. Put yourself in the other person's shoes and don't act passively by merely talking. For example, if your child reaches out to you for help on any matter, don't just sit there and give them a lesson on what and what not to do. Instead, engage actively with the problem and try to solve it together from different angles.

Optimistic

> *Be a positive anchor.*

No one wants to be anchored to a pessimistic anchor point. Be optimistic, but at the same time be reasonable. Do not daydream and mislead the vessels anchored to you. When people reach for anchors, they reach for a positive attitude. No one likes a pessimistic and negative social network anchor point.

True

> *If you are not willing to commit, do not commit.*

Be true to, and honest with, yourself and the vessel. Do not pretend to play a role that you are not willing or only have the minimum requirements to play. No one truly anchors to an unauthentic anchor. If you can't do the job or are unqualified to do it, just say so. If you're not willing to commit, do not commit, as this will save time for both you and others.

ANCHOR SYSTEM THINKING

Robust

> *Be robust, because vessels depend on you.*

An anchor point should be able to withstand changes. You are the anchor point, so you are the fixed point that vessels depend on and hold onto. As a parent, you are responsible for providing for your family—you can't just walk away when there is an economic crisis.

Under stress and in difficult times, a general cannot leave his army in the middle of the field and just resign.

Flexible

> *Give space, but always be near when needed.*

Sometimes, you'll need to be flexible with the vessels. Imagine the anchor chain is made of flexible cotton rope. This will allow enough length for the vessel to move, but still provides the care needed from your side.

At work, delegate and trust your employees—do not micromanage. At home, give your kids the freedom to choose and take responsibility, give them space, but always be there when they stumble.

Resilient and agile

> *Be creative now, for a later comeback.*

Sometimes it's better to cut the anchor chain loose for a while, but keep the vessel in your sight. So, when the right moment comes again, you can regain your anchor point position. Be creative.
Say you're working on a merger, but it seems like a current trade war would hinder the acquisition. Cut loose the chain until the right moment comes again.

Remember, just as others are anchor points for you, you are an anchor point for others, so you better do it right.

ANCHOR SYSTEM THINKING

In the following chapter, we'll explore how to use the anchor system for strategic planning.

CHAPTER 6

PROACTIVE ANCHORING

In this chapter, you'll learn how to use anchor system thinking as a strategic planning tool for yourself, your organization, and society. So instead of being reactive, you can become *proactive*. Rather than waiting to be anchored passively, why not define your anchor points in advance?

You'll start with a blank piece of paper that needs to be filled with future plans. When and where do you want to go? What and who do you want to be anchored to? Which anchor points will you choose?

In the next section, I'll show you how to use anchor maps to set personal goals, improve relationships, and do business planning.

ANCHOR SYSTEM THINKING

PERSONAL PROACTIVE ANCHOR MAP

Remember I'm planning to run a marathon next February? I'll show you how to use the anchor system and a method called **proactive anchoring** to draw an anchor map for this goal. An anchor map for your personal goal will help you visualize your goal clearly, discover your strengths and weaknesses, follow up on your plan, and tweak it along the way.

- I start by drawing myself at the center of the map.
- Then I draw arrows to the different anchor points:

Time anchor points:
- *Future time anchor point*: Next February is the race day.
- *Present time anchor point*: I write down my training plan.

- *Past time anchor point*: I write my current personal records for different distances, such as 5K, 10K, and half-marathon.

Location anchor points:
- Race city.
- Race route.
- Expected weather for race day.

Social network anchor points:
- My running group.
- My support teams such as my family, wife, and kids.
- Co-workers and friends.

Entity anchor points:
- *Objective*: My sports clothes, shorts, t-shirt, socks, and running shoes.
- *Subjective*: Ask myself, "Why do I want to finish the marathon?" A clear answer will help me through this tough journey.
- *Inter-subjective*: Write down why people run, and the physical and mental benefits of running.

ANCHOR SYSTEM THINKING

After drawing the proactive anchor map, identify which anchor points need to be strengthened to achieve your goal. Make sure you have the necessary data, such as a training program, and the necessary resources, such as running clothes.

Inform your social network anchor points about your goal to get support and understanding. For example, you'll often need to excuse yourself from social outings so you can wake up early the next morning to train.

Hang the anchor map on the wall for yourself, or you may want to share it with others.

Visit the map periodically to see how far you've got and where are you going, then update it and make changes accordingly.

A. I. Shoukry

RELATIONSHIP PROACTIVE ANCHOR MAP

The anchor system and proactive anchoring are also handy when it comes to relationships. One of the most common relationships between two people is a love relationship. There are many times when the anchor system can be helpful here. For example, if there are fights and problems when the couple is deciding whether to make a long-term commitment. Say both partners choose to get married, then they can use proactive anchoring and anchor maps to find out each other's current anchor points and define their future anchor points together.

In this anchor map example, the two partners are the vessels and the medium is their relationship. Now let's identify the anchor points they both have, currently share, or are willing to share to explore the relationship between them:

ANCHOR SYSTEM THINKING

- Identify the anchor points that hold the vessels (the couple) together, and see which anchor points they can work on together to strengthen the relationship.
- Identify which anchor points hold each vessel (partner) separately, and see which anchor points they can start to share, delete, or weaken to have a better relationship.
- Identify what new anchor points they can create to hold both vessels together in a stronger way.
- Explore all time, location, social network, and entity anchor points.
- Draw a new anchor map that they share together.

ORGANIZATIONAL PROACTIVE ANCHOR MAP

The anchor system and proactive anchoring can also be used to draw anchor maps when you're planning at the organizational level.

For example, say you're starting a new company, and you need to define its future anchor points.

Time anchor points:
- *Past time anchor point:*
    - Founders history
    - Brand history
    - Product history
- *Present time anchor point:*
    - One-year plan
    - Vision and mission
- *Future time anchor point:*
    - Future plan

Location anchor points:
- Location of the company

ANCHOR SYSTEM THINKING

- Target market
- Distributors
- Future store location

Social network anchor points:

- Key people in the company
- Key relations outside the company

Entities anchor points:

- *Objective entities:*
    - Resources
- *Subjective entities:*
    - Company culture
- *Inter-subjective entities:*
    - Market
    - Money value
    - Country's law
    - The moral code of the company
    - Values

By having an anchor map for your new business, you can:

Easily spot the strong and weak points.

A. I. Shoukry

Have a clear map of your current and future social networks.

Identify where your company is heading for both yourself and your workers.

ANCHOR SYSTEM THINKING

SOCIETY PROACTIVE ANCHOR MAP

The anchor system can also be used for marketing and political campaigns, and in many other applications that involve planning and decision-making. Here is an example of a candidate preparing for an election campaign:

Vessel:
- The candidate
- The voters

Medium:
- The election campaign

Anchor Map:
- Identify the anchor points that both the candidate and the voters hold.
- Identify which anchor points hold the voters to the candidate, and therefore can be enforced and highlighted.
- Identify which anchor points the candidate holds but that alienate a segment of voters and therefore can be weakened or deleted.

- Identify what new anchor points the candidate should create that are shared with voters.
- Explore all time, location, social network, and entity anchor points.
- Draw an anchor map for the campaign.

CONCLUSION

In this short book, you have been introduced to anchor system thinking and now know how to use it to anchor in a sometimes stressful and chaotic world. You have now mastered the art of situational analysis using anchor system thinking problem solving, and strategic planning for your future, your organization, and society,

At the end of the book, I'll leave you with some mind gymnastics games. If we are the vessels, life is the sea, and we are anchored by anchor points, what are the ports and port services? How would you handle the turbulent, high seas and what would you do on the calm days?

In the Appendix, you'll find the complete anchor system with its detailed parameters. Adding this level of detail will help you take the system even further.

Anchor system thinking will change the way you think, analyze, and plan for yourself, your organization, and society forever.

Happy anchoring!

APPENDIX

ANCHOR SYSTEM

Being anchored requires five components:

- The vessel
- The medium
- The anchor
- The bed
- The chain

The vessel is tied with a chain and an anchor to the bed of the medium. Together, these five components form the anchor system.

The anchor (A), bed (B), and chain (C) are the ABC of the anchor point. This is what each component means.

The anchor

The anchor is the hook that connects to a vessel via a chain on one end, and hooks to the bed of the medium on the other end. Anchors can be permanent or temporary.

The bed

An anchor hooks a vessel to the bed of the medium. The anchor bed can be fixed or dynamic. A fixed bed means the vessel decides when to anchor to the bed and when to be free. A dynamic bed is where the anchoring is a two-way relationship, and either side can remove the anchor if they wish to.

The chain

The chain holds the vessel to the anchor can be made of steel, which is hard to break, or a rope, which is made of textile and is much easier to break. The chain may be of fixed length or flexible, which allows some maneuvering.

The vessel

The vessel can be a person such as yourself, friend, husband, wife, etc. Or it can be an organization such as a company, charity organization, political party, etc. It can also be a nation, country, or religion, etc.

The medium

The medium can be life in general, marriage in relationships, a political sphere for political parties, the market for companies, geopolitics for countries, society for religion, etc. The medium may be regulated or unregulated, slow-moving, fast-moving, or stable (idle).

A. I. Shoukry

ACKNOWLEDGEMENTS

First, I would like to thank my primary anchor, my wife Heba, for always being there for me.

Special thanks to Brian Baker for his editorial assessment and to my newsletter members who were the first draft readers, especially Khalid Hussein.

Lastly, I would like to thank my editor Ameesha Green. I thank her for her continuous support as we celebrate our third collaboration together.

For icons featured in the anchor map key, credit goes to:

People Icon made by Smashicons from www.flaticon.com

ANCHOR SYSTEM THINKING

Social Network Icon made by Smashicons from www.flaticon.com

Time Icon made by Icon Works from www.flaticon.com

Location Icon made by Freepic from www.flaticon.com

Entity Icon made by Freepic from www.flaticon.com

A. I. Shoukry

ABOUT THE AUTHOR

Ahmed I. Shoukry is an Associate Professor of Urology at Cairo University, Egypt. He started blogging in 2004 and launched a portal for independent bloggers in 2007. In 2018 he released a bestselling book titled "*In or Out: A Practical Guide to Decision Making*" and a running memoir titled "*It's Not Just About Running: Reflections on Life and Change in Egypt*."

Twitter: @ashoukry

Facebook: Ahmed.Ismail.Shoukry

Website: Shoukry.org

Instagram: @AhmedIShoukry

To get exclusive content and free materials subscribe to A. I. Shoukry Newsletter at Shoukry.org/Newsletter.

REFERENCES

[1] Anchor | Definition of Anchor by Merriam-Webster n.d. https://www.merriam-webster.com/dictionary/anchor (accessed March 29, 2018).

[2] anchor | Definition of anchor in English by Oxford Dictionaries n.d. https://en.oxforddictionaries.com/definition/anchor (accessed March 30, 2018).

[3] Anchor - Wikipedia n.d. https://en.wikipedia.org/wiki/Anchor (accessed March 30, 2018).

[4] Our mission and vision | Volvo Group n.d. https://www.volvogroup.com/en-en/about-us/our-mission-and-vision.html (accessed

August 17, 2018).

[5] As Amazon narrows choice for HQ2, these cities finishing strong n.d. https://www.cnbc.com/2018/08/16/as-amazon-narrows-choice-for-hq2these-cities-finishing-strong.html (accessed August 17, 2018).

[6] SCA - Canal History n.d. https://www.suezcanal.gov.eg/English/About/SuezCanal/Pages/CanalHistory.aspx (accessed August 17, 2018).

www.ingramcontent.com/pod-product-compliance
Lightning Source LLC
Chambersburg PA
CBHW020448220526
45464CB00002B/905